Crooked Cracked Tree in Fog

Todd Hido

Crooked Cracked Tree in Fog

Nazraeli Press

Brian S. Sancha-Dorantes

BY

Cindy Patino

Cindy Patine

Edwin Felipe-Pozo

Nayra Y. Rodriguez

Stop
Nayra

Crooked Cracked Tree in Fog
A book of four reproductions of collages made by students in Ms. Campos' 08-09 bilingual 1st grade at Lyon Magnet Elementary School, Waukegan, Illinois under Mr. Matthies' direction, and one original photograph by Todd Hido.

N° 339/500 (Print signed on verso)

Nazraeli Press One Picture Book N° 60
Printed in an edition of 500 copies

Order Address: Nazraeli Press
2871 NE Alameda Street
Portland, Oregon 97212, USA

Printed in China for Nazraeli Press
ISBN 978-1-59005-292-1